MATH TO MUNCH ON

BY MARC GAVE

Table of Contents

Growing Your Body

Chocolate is OK to eat on special occasions.

A small amount of broccoli contains a lot of the vitamins and minerals that keep you healthy.

Have you ever heard the word **nutrition** (noo-TRIH-shun)? What does it make you think of? Passing up chocolate cake for an apple? Forcing yourself to eat broccoli even though you don't like it?

Nutrition is all about food. It's about how food provides energy and helps the body grow.

Every time you eat, you make choices between foods that are good for your body and foods that are not so good for your body. This book will help you choose foods that make your body healthy and strong. After all, you eat a lot of meals in your lifetime.

SOLVE THIS!

1. If you eat three meals a day, how many meals do you eat
   - in a week?
   - in a month?
   - in a year?
2. If you live for 80 more years, how many meals will you have eaten in your lifetime? Use a calculator to find out.

Food for Thought

The word **diet** refers to what you eat. It is very important to have a balanced diet. That means you should eat many different kinds of foods. For example, you should eat foods that are rich in **protein** (PROH-teen), such as meat, milk, nuts, and beans. You should also eat foods that are rich in **carbohydrates** (kar-boh-HYE-drayts), such as cereal, bread, noodles, fruits, and vegetables. And your body also needs a certain amount of **fat**. Fats are found in foods such as eggs and milk.

These people are taking part in an eating contest. They are *not* eating a balanced diet. Fortunately, they don't eat like this every day.

FOODS RICH IN PROTEIN (meat, milk, nuts, beans)

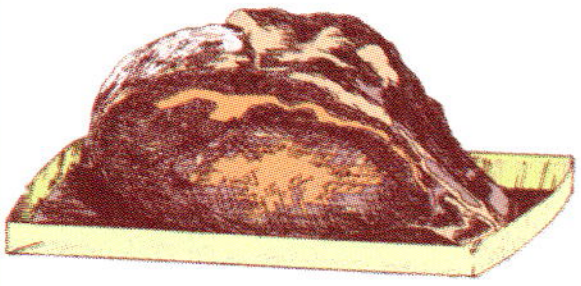

FOODS RICH IN CARBOHYDRATES (cereal, bread, noodles, fruits, vegetables)

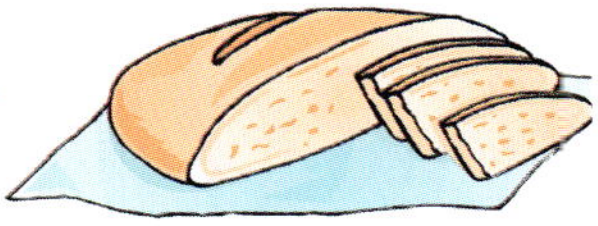

FOODS RICH IN FAT (eggs, milk)

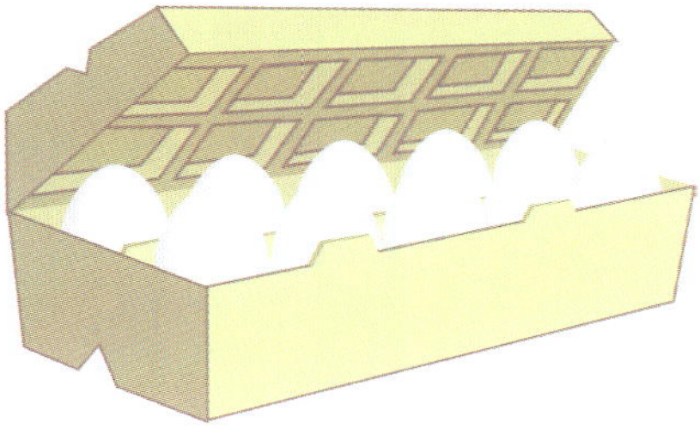

Diet

Sometimes the word "diet" is used to mean what you should or shouldn't eat if you want to gain or lose weight. If you have any questions about your diet or weight, talk them over with a doctor or an adult whom you trust.

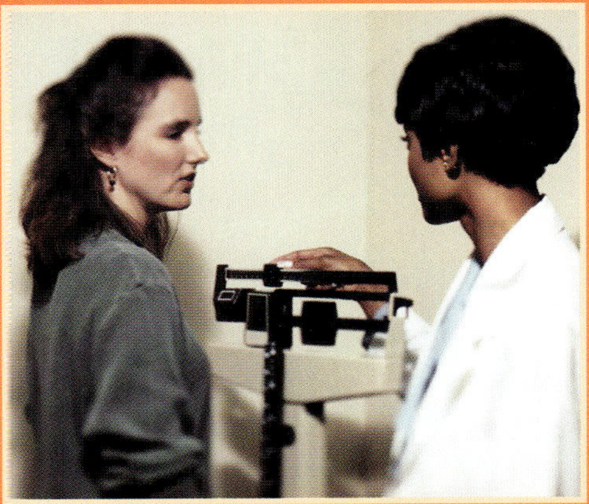

It is important to speak to someone about a healthy diet and weight for you.

To guide you in eating a balanced diet, the U.S. government has created the Food Guide Pyramid. It's probably the best guideline you can use to eat well.

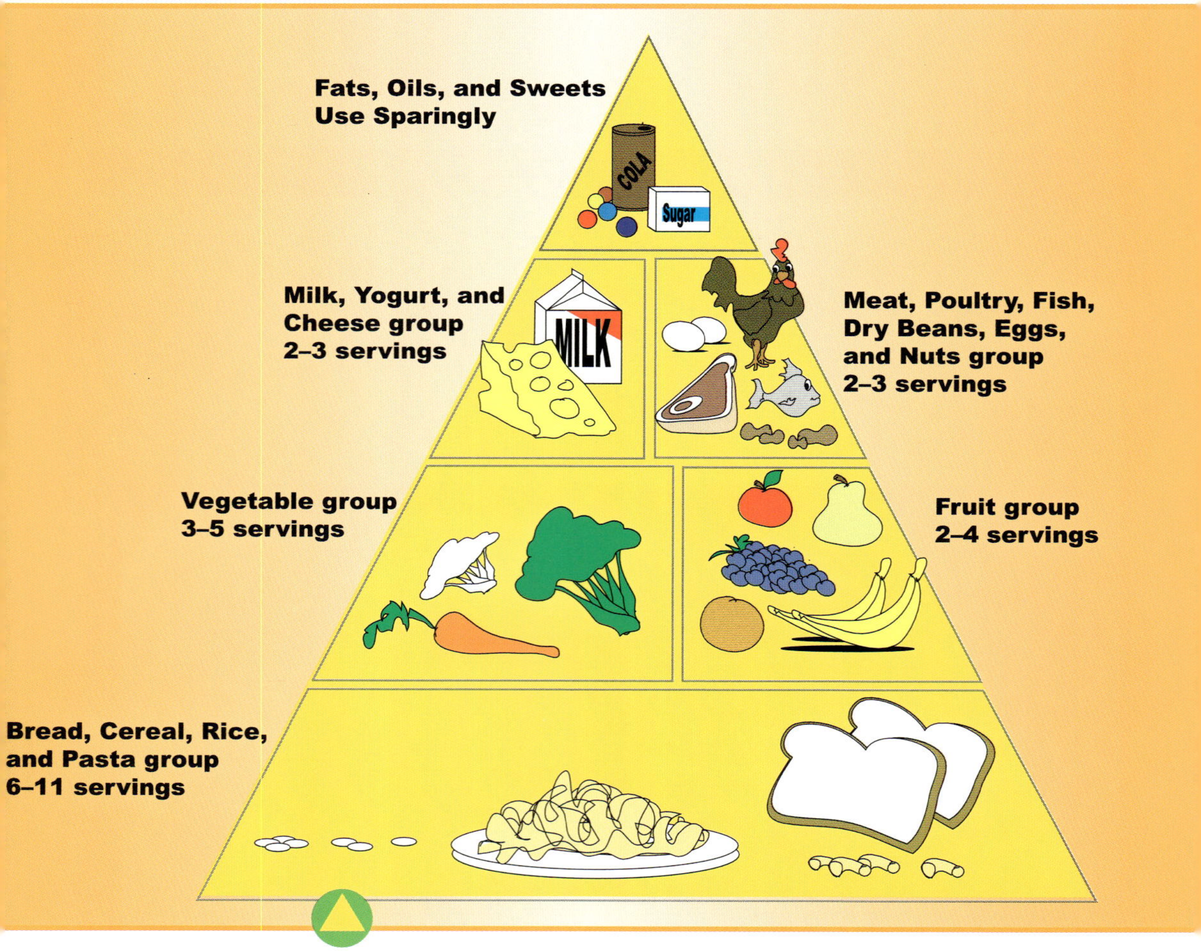

This shows how many servings a person should eat from each food group. Notice that the foods at the bottom of the pyramid should be eaten more often than the foods at the top.

It's a FACT!

Foods contain vitamins and minerals that your body needs. By eating a certain number of servings a day of different foods, you can make sure that you are getting enough vitamins and minerals. Your doctor may suggest taking a vitamin pill every day. That gives your body a boost, but it doesn't take the place of a healthful diet.

What Counts as a Serving?

Food Groups			
Bread, Cereal, Rice, and Pasta	1 slice of bread	1 ounce of ready-to-eat cereal	1/2 cup of cooked cereal, rice, or pasta
Vegetable	1 cup of raw vegetable	1/2 cup of other vegetable, canned, cooked, chopped	3/4 cup of vegetable juice
Fruit	1 medium apple, banana, or orange	1/2 cup of chopped, cooked, or canned fruit	3/4 cup of fruit juice
Milk, Yogurt, and Cheese	1 cup of milk or yogurt	1 1/2 ounces of natural cheese	2 ounces of processed cheese
Meat	2 to 3 ounces of cooked lean meat, poultry, or fish	1/2 cup of cooked dry beans or one egg counts as 1 ounce of lean meat. 2 tablespoons of peanut butter or 1/3 cup of nuts counts as 1 ounce of meat	

Here are serving sizes for foods in the five food groups.

In its Food Guide Pyramid, the U.S. government measures food in terms of servings. But what exactly is a serving? That depends on the type of food you are eating.

Food amounts are usually measured by the **gram** or by the **ounce**. A gram (g) is a unit of mass in the metric system. An ounce is a unit of weight in the system we use in the United States. Food packages usually list the weight of the food in grams and in ounces. One slice of white bread has a mass of about 25 grams and weighs a little less than one ounce.

The amount of energy in food is measured by the **calorie** (KAL-uh-ree). For example, a slice of white bread has 65 calories.

BREAD

One 25-gram slice of white bread provides about 65 calories of energy.

Why would the slice of bread shown here have more than 65 calories?

SOLVE THIS!

3 **If there are about 28 grams in an ounce and 16 ounces in a pound, how many grams are there in a pound? Use a calculator to find out.**

What about food that isn't sold in packages? Do you have to measure everything before you put it into your mouth? Not really. Nutrition books and cookbooks often have tables that list how many calories and grams are in foods. Even some fast-food restaurants list the number of calories in their food.

It's a FACT!

For every 100 calories you eat, you should try to get no more than 30 of them from fat.

Nutrition Facts

Serving Size 1 cup (30g)
Children Under 4 - ¾ cup (20g)
Servings Per Container About 9
Children Under 4 - About 14

Amount Per Serving	Cereal	with ½ cup skim milk	Cereal for Children Under 4
Calories	110	150	70
Calories from Fat	15	20	10
	% Daily Value**		
Total Fat 2g*	**3%**	**3%**	1g
Saturated Fat 0g	**0%**	**3%**	0g
Polyunsaturated Fat 0.5g			0g
Monounsaturated Fat 0.5g			0g
Cholesterol 0mg	**0%**	**1%**	0mg
Sodium 280mg	**12%**	**15%**	190mg
Potassium 95mg	**3%**	**9%**	65mg
Total Carbohydrate 22g	**7%**	**9%**	15g
Dietary Fiber 3g	**11%**	**11%**	2g
Soluble Fiber 1g			0g
Sugars 1g			1g
Other Carbohydrate 18g			12g
Protein 3g			2g
			% Daily Value
Protein	-	-	9%
Vitamin A	10%	15%	10%
Vitamin C	10%	10%	10%
Calcium	10%	25%	8%
Iron	45%	45%	50%
Vitamin D	10%	25%	6%
Thiamin	25%	30%	35%
Riboflavin	25%	35%	35%
Niacin	25%	25%	35%
Vitamin B_6	25%	25%	45%
Folic Acid	50%	50%	60%
Vitamin B_{12}	25%	35%	30%
Phosphorus	10%	25%	8%
Magnesium	10%	10%	10%
Zinc	25%	30%	30%
Copper	2%	2%	2%

*Amount in Cereal. A serving of cereal plus skim milk provides 2g total fat (0.5g saturated fat, 1g monounsaturated fat), less than 5mg cholesterol, 350mg sodium, 300mg potassium, 28g total carbohydrate (7g sugars) and 7g protein.

**Percent Daily Values are based on a 2,000 calorie diet. Your daily values may be higher or lower depending on your calorie needs:

	Calories:	2,000	2,500
Total Fat	Less than	65g	80g
Saturated Fat	Less than	20g	25g
Cholesterol	Less than	300mg	300mg
Sodium	Less than	2,400mg	2,400mg
Potassium		3,500mg	3,500mg
Total Carbohydrate		300g	375g
Dietary Fiber		25g	30g

A food label such as this contains important information about the nutritional value of food. You can use the information to help plan and eat a balanced diet.

Eat Hearty!

SOLVE THIS!

4 **Add up the grams of fat and the calories that Michael and Marisa eat for each meal.**

Now it's time to put into practice what you've learned.

Michael and Marisa are brother and sister. They eat the same things for breakfast, lunch, and dinner. They get their choice of an afternoon snack and dessert after dinner. In the following three charts, you'll see what they ate for their three meals.

BREAKFAST

	Amount	Fat (g)	Calories
Orange juice	1 cup	0	110
Scrambled eggs	2	14	200
Margarine, low-calorie	1 tbsp.	5	50
Milk, skim	1 cup	0	90
Toast	2 slices	2	130
TOTAL for Breakfast			

LUNCH

	Amount	Fat (g)	Calories
Peanut butter and jelly sandwich	1	18	375
Celery	1 stalk	0	5
Carrot	1	0	30
Milk, skim	1 cup	0	90
Apple	1	0	80

TOTAL for Lunch

DINNER

	Amount	Fat (g)	Calories
Chicken-noodle soup	1 cup	6	175
Chicken drumstick, roasted	1 piece	6	112
Rice	1 cup	0	205
String beans	1 cup	8	150
Hard roll	1	2	167
Water	2 oz.	0	0

TOTAL for Dinner

	Fat (g)	Calories
Breakfast	______	______
Lunch	______	______
Dinner	______	______
TOTAL	______	______

SOLVE THIS!

5 Now total the grams of fat and the calories that Michael and Marisa have eaten at their three meals.

6 If Michael and Marisa should eat no more than 73 grams of fat and 2,200 calories, how many grams of each do they have left for their afternoon snack and evening dessert?

SOLVE THIS!

7 What are some foods Michael and Marisa might choose for snack and dessert that would not put them over their daily totals of fat and calories?

SNACK & DESSERT FOODS	Amount	Fat (g)	Calories
Apple pie	1 slice	18	405
Applesauce, sweetened	1 cup	0	195
Banana	1	0	105
Brownie with nuts	1	4	100
Chocolate-chip cookies	4	9	180
Ice cream, low-fat	1 cup	6	185
Ice cream, vanilla	1 cup	14	270
Popcorn, no butter	1 cup	0	30
Potato chips	10	7	105
Pretzels, thin	5	1	120
Pudding, chocolate	½ cup	4	155
Raisins	1 packet	0	40
Yogurt, low-fat, fruit	8 oz.	2	230

CHAPTER 4

What's Cooking?

Another way to learn how to eat well is by helping to plan your meals. Helping to shop for your food and cook it also teaches you good eating habits.

Some supermarkets collect packaged food to give to people in need. Does your school have a food drive?

When it comes to cooking, one of the first things you need to know is how to follow a **recipe**. If you use a recipe, you know that what you are cooking has worked for other cooks.

SOLVE THIS!

8 How many times more is 1 teaspoon than
- 1/2 teaspoon?
- 1/4 teaspoon?
- 1/8 teaspoon?

Recipes require measuring. Cups and spoons are the measuring tools most often used in the kitchen. These are not the kinds of cups you drink out of or the kinds of spoons you use to eat ice cream or soup. These are special ones that are designed for measuring.

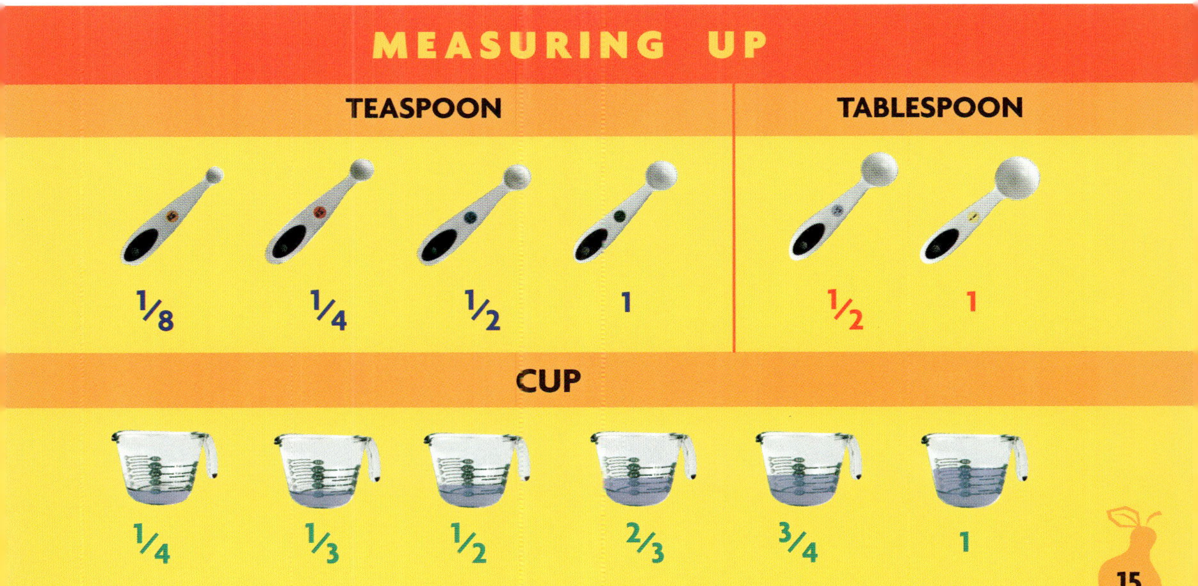

Counting Up

It's important to know how teaspoons, tablespoons, and cups are related.

1 tablespoon (tbsp.) = 3 teaspoons (tsp.)
1 cup (c.) = 16 tablespoons (tbsp.)

SOLVE THIS!

9. How many tablespoons are in
   - 1/2 cup?
   - 1/4 cup?

10. How many teaspoons are in 1 cup?

11. How many teaspoons are in 1/4 cup?

Two other measurements we use in the kitchen are time and temperature. Most ovens have a timer that you can set. If yours does not, you can use a separate timer or an alarm clock.

If you're baking something in the oven, it is a good idea to check it a few minutes before it is supposed to be done. Oven temperatures are sometimes not exact. Food may take more or less time to cook than you expect.

SOLVE THIS!

12 Some recipes tell you how much time it takes to prepare the food as well as how much time it takes to cook it. If it takes 35 minutes to prepare a recipe and then 1 hour and 15 minutes to cook it, how much time will it take altogether?

Most ovens have a measuring dial for setting the temperatures.

If an oven doesn't have a timer, you can use a separate one.

Banana Bread

Here's an easy recipe for you to try.
Have an adult help you.

Here's what you need to make a loaf:

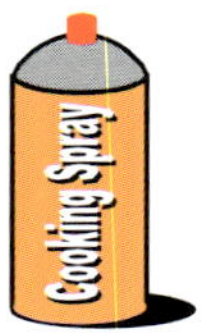

Cooking spray

1/2 cup milk

2 eggs

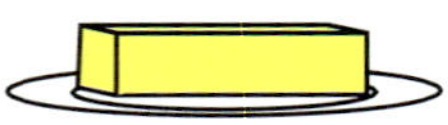

4 tbsp. margarine, softened

1 cup sugar

1 tsp. vanilla

1 1/4 cup mashed bananas (about 2 large bananas)

1 3/4 cup flour

1 tsp. baking soda

1 tsp. cinnamon

1/2 tsp. salt

1/4 tsp. baking powder

Banana Bread

Here's what you do:

- Have an adult heat the oven to 350° Fahrenheit.
- Spray an $8\frac{1}{2} \times 4\frac{1}{2} \times 2\frac{1}{2}$-inch loaf pan with cooking spray.
- In a large bowl, place milk, eggs, margarine, sugar, and vanilla. Have an adult beat the mixture for 30 seconds with an electric mixer.
- Add mashed bananas and beat at high speed for a minute longer.
- In another bowl, place flour, baking soda, cinnamon, salt, and baking powder. Mix with a large spoon.
- Add flour mixture to the wet ingredients and mix until blended.
- Spread mixture evenly in the loaf pan. Bake until the bread is golden brown. That should be about 60 minutes. Bake longer if you need to. To test if it is done, stick a toothpick into the center. If the bread is done, the toothpick will come out clean.
- Let bread sit in the pan on a wire rack for five minutes. Then remove the loaf from the pan and cool it on the rack.

SOLVE THIS!

13 How much of each ingredient would you need to make two loaves?

It's a FACT!

In the United States, temperature is measured in degrees Fahrenheit. In most other countries, it is measured in degrees Celsius.

To change Fahrenheit into Celsius:

Take the degrees Fahrenheit and subtract 32. Multiply that by 5 and divide the product by 9.

The banana bread recipe calls for baking at 350° Fahrenheit for 60 minutes. Could you bake it at 175° Fahrenheit for two hours instead? No. You would end up with a soggy mess. And if you have only 30 minutes, leave baking for another time. Even if home ovens went up to 700° Fahrenheit, baking it at that temperature wouldn't work.

Using mitts or pot holders keeps your hands safe when handling hot foods.

The Missing Ingredient

These skaters know that wearing safety equipment is very important.

Eating well is not the only thing that keeps you healthy. It is also important to have regular checkups with your doctor and dentist. And it is important to exercise.

Exercise keeps your body in top shape. What's your favorite way of exercising?

No one loves getting a dental checkup. But most dentists try to make it fun.

Solve This! Answers

1. page 3
   21; around 90; 1,095
2. page 3
   87,600, plus whatever you have already eaten
3. page 8
   around 448 g
4. page 10–11
   Breakfast totals: 21 g fat, 580 calories
   Lunch totals: 18 g fat, 580 calories
   Dinner totals: 22 g fat, 809 calories
5. page 12
   Daily totals: 61 g fat, 1,969 calories
6. page 12
   Still can eat: 12 g fat, 231 calories
7. page 13
   Answers will vary.
   Possible answers:
   Snack: pretzels
   Dessert: banana—total 1 g fat, 225 calories
   Snack: popcorn
   Dessert: chocolate pudding—total 4 g fat, 185 calories
8. page 15
   2 times; 4 times; 8 times
9. page 16
   8; 4
10. page 16
    48
11. page 16
    12
12. page 17
    1 hour, 50 minutes
13. page 19
    1 c. milk; 4 eggs; 8 tbsp. margarine; 2 c. sugar;
    2 tsp. vanilla;
    $2\frac{1}{2}$ c. mashed bananas;
    $3\frac{1}{2}$ c. flour;
    2 tsp. baking soda;
    2 tsp. cinnamon;
    1 tsp. salt;
    $\frac{1}{2}$ tsp. baking powder

Glossary

calorie	unit of heat contained in food
carbohydrates	chemical substances found in plant foods, such as sugars and starches
diet	what you eat
fats	chemical substances found in meat and in the nuts, seeds, and fruits of plants
gram	unit of mass in the metric system
nutrition	the food you eat; eating food for energy and growth; the study of food
ounce	unit of weight
protein	a chemical substance necessary for all living cells; contained in such foods as meat, fish, eggs, milk, and beans
recipe	directions for preparing food

Index